Lizzie Penney

The Juvenile Temperance Reciter

Lizzie Penney

The Juvenile Temperance Reciter

ISBN/EAN: 9783337371593

Printed in Europe, USA, Canada, Australia, Japan

Cover: Foto ©berggeist007 / pixelio.de

More available books at **www.hansebooks.com**

THE

Juvenile Temperance Reciter

No. 5.

A COLLECTION

OF

CHOICE RECITATIONS AND DECLAMATIONS,

IN PROSE AND VERSE,

FOR USE IN

**SABBATH-SCHOOLS, DAY-SCHOOLS, BANDS OF HOPE, JUVE-
NILE TEMPLES, LOYAL TEMPERANCE LEGIONS, JUNIOR
SOCIETIES OF CHRISTIAN ENDEAVOR, AND ALL
JUVENILE ORGANIZATIONS.**

EDITED BY

Miss L. PENNEY,

Editor of "The Juvenile Temperance Reciter," Nos. 1, 2, 3, and 4: "Concert
Temperance Exercises, or Helps for Entertainments," "How to Fight
the Drink, or the Saloon Must Go." etc.

NEW YORK:

The National Temperance Society and Publication House,

58 READE STREET.

PREFACE.

BECAUSE there is a constant demand for new material for use in Children's Organizations, and especially in the line of entertainment, this collection of recitations and declamations has been made, by request.

While some of its contents were culled from various sources, the greater portion is new and fresh, prepared especially for the Compiler.

We tender our sincere thanks to Mr. Edward Carswell and Mr. George W. Bain, the famous speakers, and to other kind friends who responded promptly to our request for recitations.

THE COMPILER.

CONTENTS.

PROSE.

POETRY.

(5)

The New and Political Girl.

EDWARD CARSWELL.

[This should be assigned to a good-sized Girl, who must speak slowly and with animation.]

THAT the new woman's come, of course you have heard,
  For she came with a rush and a whirl;
But you may not have seen or heard, until now,
  A new and political girl.
So I'll make you a speech, but not on "our rights,"
  Nor "the tariff," or "sixteen to one";
It's an issue that far transcends any of these,
  And *something* has *got* to be done!
We are spending three thousand a minute for drink,
  Day and night, year out and year in;
And our health, wealth, and freedom are drifting away
  On this black foaming river of sin.
Such questions as "poverty," "taxes," and "trusts,"
  We have shouted, and spouted, and sung,
And have tried at the spigot the driblets to stop,
  While our wealth disappears at the bung.
Three thousand a minute! Just think of it, friends,
  And this but the mere money cost;
Not counting the shame, the sorrows and crimes,
  Nor the lives and the souls which are lost.

(7)

Three thousand a minute! Oh, what it would do
  If used in a sensible way!
We could start the most wonderful boom in our land
  With this money we're throwing away.
No man need be idle, no home without bread;
  It would light up each factory and mill;
And industry's wheels be heard humming again
  Where now they are silent and still.
Then arouse on *this* issue (the greatest of all),
  Let the minor ones rest on their shelves;
And when we have settled this question of drink,
  The others will settle themselves.
Then marshal your forces against the saloon;
  Prohibition's white banner unfurl;
Destroy this foul monster! and also the need
  Of a New and *Political* Girl.

Complaints.

MRS. VIRGINIA J. KENT.

[This can be recited by one Child, or divided among five.]

"O my! O my!" said the pretty rye,
"I feel bad enough to have a good cry.
I thought I was meant to be used for food,
And was planted and grown to do some good,
But now, when I've done my best, just think!
I'm converted into a nasty drink.
If I had known about this last fall,
I really don't think I'd have grown at all."

A stalk of corn bowed its graceful head
And sighed, "I almost wish I were dead!
For the same disgraceful fate, I fear,
Awaits my every ripening ear.

'Tis a burning shame to use us so,
And force us, for such a base purpose to grow."

A murmur arose on the summer air,
A murmur of sorrow, and grief and despair,
Among the hop-vines, as they trembled with fear,
For they knew they were doomed to make ale and beer;
And they mourned that in all their beauty and pride
They must be on the brewers' and drunkards' side.

Then the grape vines and apple trees looked around,
And shook from their topmost leaves to the ground,
As the farmer talked of the cider and wine
He should make in the fall from each tree and vine.

Let us pray and work for the happy day,
When temperance folks can have their way,
All through our land, from east to west
(For temperance folks know what is best);
And by the time that we are all grown
'Twill be the best country that ever was known.

Protect the Boys.

THOS. R. THOMPSON.

[For a small Boy.]

I'M the little boy
    Who would like to see
Every land and State
    From the drink curse free.
And I want to give
    Just a thought or two
As to what I think
    Every one should do.

Every one should look
  At the now—and then—
For we boys, you know,
  Are the coming men.
Every boy should strive
  With his utmost might
To adhere to that
  Which is good and right.

And you men should help;
  For the boys need care.
Not a boy now born
  Can the nation spare.
Then the nation's curse
  From our presence take.
Won't you do it, please,
  For the children's sake?

———

" If."

WILLIAM HOWARD.

[For either a Girl or Boy.]

IF you want a red nose and dim, bleary eyes;
If you wish to be one whom all men despise;
If you wish to be ragged and weary and sad;
If you wish, in a word, to go to the bad;
          Then drink!

If you wish that your life a failure may be;
If you wish to be penniless—out at the knee;
If you wish to be houseless, broken, forlorn;
If you wish to see pointed the finger of scorn;
          Then drink!

If you wish that your manhood be shorn of its strength;
That your days may be shortened to one-half their length;
If you like the gay music of curse or of wail;
If you long for the shelter of poorhouse or jail;
      Then drink!

If your tastes don't agree with the "ifs" as above;
If you'd rather have life full of brightness and love;
If you care not to venture nor find out too soon
That the gateway to ruin lies through the saloon!
      Then don't drink!

I'm Only Six.

W. A. EATON.

[For either a small Girl or Boy.]

I'M only six years old, you know,
 But my big brother John
Said I must say a little piece
 To show which side I'm on.

So I have come to say to all
 Who hear me speak to-night,
That you can see which side I'm on;
 Of course, I'm on the right.

The birds that sing so sweet and clear
 Drink water pure and bright,
And that is best for you and me;
 That's all I'll say to-night.

What I Think.

IF drinking liquor often leads
 To sorrow and to sin,
Then every boy and girl should say,
 "I never will begin!"

Sammy's Idea of the Silver Question.

MRS. M. A. KIDDER.

[For a Boy.]

LITTLE bits of silver,
    Little grains of gold,
Make a lot of trouble—
    How the people scold!
If we only had some—
    Just a few, you see—
We'd know how to spend 'em,
    Johnny Bates and me!

First we'd buy some candy—
    Half a pound, or more;
Then we'd buy some cookies
    Down at Kneadham's store.
When we'd done a-feasting,
    In some shady nook,
Then we'd wash it down with
    Water from the brook.

Little grains of silver,
    Little chunks of gold,
Make a lot of trouble
    Where strong drink is sold.
But *we'll* never buy it—
    Just you wait and see—
Rum or gin or brandy,
    Johnny Bates and me!

A Little Boy's Resolve.

I WILL never use tobacco!
    I will never swear; no, never!
I'll drink no wine nor whiskey!
    I'm a teetotaler forever!

The Screen-door.

HELEN E. BROWN.

[For a Boy.]

" Mamma, there's something I'd like to know."
      Said Archie McKee one day;
" As up and down the streets I go,
      On either side of the way,
   I notice at every single store—
      Where the sign says whiskey and gin—
   There's a curious-looking swinging door,
      So I can't get a look within.

" I've tried and tried to catch a sight
      Of what goes on inside:
   The men go in, and the door shuts tight,
      As if there was something to hide.
   At the stores where clothing is sold, and meat,
      And candy, and bread, and shoes,
   The doors are open on the street,
      We can look in as much as we choose.

" Now, mamma, what is it they do in there
      Where the whiskey and gin are sold,
   That needs such very particular care
      Lest the folks outside behold?
   They drink, I suppose; but, if it's right,
      Why, then, should they be afraid?
   Why don't they do it in open sight,
      And not behind a shade?"

" It's true, my son, we need not fear
      If we know we are doing right;
   We can drink our whiskey, cider, or beer
      All out in the broad daylight.

But you see the screen-doors signify,
   Wherever they are hung,
That the men who sell and the men who buy
   Are doing a fearful wrong.

" And more, they know they are doing ill,
   Which makes it the greater sin ;
Ah ! deep and dark the record they fill
   Inside the bar-room screen.
They keep their doings from sight of men,
   But they can do no more ;
God sees as clearly behind the screen
   As if 'twere an open door."

Then said the boy, " Whene'er I see
   That curious swinging screen
I'll know that dark iniquity
   Is going on within ;
And I'll set myself with a mighty will
   Against the liquor-stores—
Against the bars—against the still—
   And the evil behind the doors.

" But I'll try to save the wretched men
   Who are enticed within—
Like flies into the spider's den—
   To perish in their sin.
As long as I have life and breath,
   That life and breath I'll spend
To snatch the tempted ones from death—
   The drunkard's fearful end."

What a Little Girl can do for Temperance.

JESSIE FORSYTH.

[For a Girl.]

OUR superintendent told us the other day that we must all try to do something for temperance, and I thought about what she said all the way home. Then I met old Mr. McFogie and I told him about it. But he said, "Oh, pshaw, what can a mite like you do?" He is not really cross, you know, and he likes to hear about our meetings, but he makes fun sometimes. And this time I felt real discouraged, especially when he said, "There, run along and take care of your dolly; that's about as much as you can do, I guess." I talked it over with Evelina (she's my doll, and she's just as sweet as can be), and I said, "Now Evelina, *my* mamma takes care of me and the rest of us, but she finds time to do a great deal for temperance, too. Don't you think that your mamma can take care of you and do a little something besides?" And Evelina looked as if she agreed with all I said.

Then I talked to mamma about it, and when I told her what old Mr. McFogie said, she just laughed and said, "Ah, that's what they always say about us women." I knew what she meant by that, because once when she told Mr. McFogie that she was just as well able to vote as any man, he said that it was a woman's business to stay at home and take care of her family.

Well, then I asked papa what I could do for temperance, but he was very busy and he only said, "Oh, you can talk to people." So, when one of mamma's friends called, I said to her, "Miss Slowpace, are you a temperance woman?" And she said, "Bless the child, what a question!" And mamma said it was quite to the point. Then Miss Slowpace told me that she did not think she ought to give up her glass of wine because she knew it did her good. I was so surprised that a grown-up lady should think such a foolish thing as that, that I forgot I was only a little girl, and I said, "Oh, that's a dreadful mistake." And then I went on and told her some of the lessons I had learned in our Temple, about the poison in wine, and how much harm even one glass would do. Mamma never scolded me for talking so much, and Miss Slowpace listened quite attentively. When I stopped, she asked me why I need trouble about the matter when nobody in my family drinks? Now wasn't that a funny question? I guess we don't belong to a Juvenile Temple just to do ourselves good. Why, one of our passwords was "work for others," and that's what we want to do. So I had to talk some more and tell Miss Slowpace that there are a great many people who are slaves to drink and that we ought to help to set them free, just as the good men and women worked to free the black slaves long ago. And mamma said something about "remembering those in bonds as bound with them," and when Miss Slowpace looked at me I could see tears in her eyes, but she only said, "I guess you are in the right, child."

To-day mamma told papa that Miss Slowpace was getting to be quite interested in the temperance cause, and that she said to tell me that she was going to help free the slaves. And papa said I was a "good little missionary," and he was not making fun, either. So I think I have found out that a little girl can do something for temperance.

Two Offerings.

MRS. O. W. SCOTT.

[For a little Girl.]

I DIDN'T think I could do it
When first he told me to,
For I love my precious dolly,
And she is almost new.

But, dear me, Uncle Joe knows how
  To talk until you feel
As if you'd give your money and
  A part of every meal.

He knows about the Jews, you see,
  And how they brought the Lord
The first and best of all their fruits
  According to His Word.
That must have been so beautiful—
  Those harvest offerings!
Well, Uncle Joe, he talked until
  I brought him all my things,
To see which I would send away
  To China in the box.
And *he* said *my best doll ;*—blue-eyed,
  Red-cheeked, with curling locks.

I said : " Do *you* give what you like,
  The very bestest best?
And do *you* ' make a sacrifice'
  As you tell all the rest?"
And he said " Yes," he always gave
  To help along the cause,
But as he had no fields or fruits
  He couldn't keep *Jewish* laws.

Now Uncle Joe is very good,
  But he does love cigars!
He smokes on the piazza till
  He almost hides the stars.
So then I said : " If you'll give up
  Cigars and pipes and all,
And give the money to the Lord,
  Why, then, I'll send my doll!"

Then Uncle Joe looked sober, for,
    You see, he loved them so;
I said: "Oh, now you see what 'tis
    To let my dolly go!"
I thought he would not do it,
    But by and by he said:
"I think you're right.   I'll drop cigars
    And give their cost instead!"

So now my dolly's going,
    And Uncle Joe—just hear!—
Will give most seventy dollars
    To missions every year!
And mamma says she's very glad
    About the way I spoke,
Since Uncle Joe has offered up
    His sacrifice of smoke!

A Voice from the Juvenile Ranks.

PETER T. WINSKILL.

[For a Boy.]

SOME say teetotalers go too far,
    And ne'er will gain their end;
Although they labor hard and long,
    Much time and money spend.
'Tis folly e'er to hope to see
    A day in this land when
The liquor-shops shall all be closed,—
    But, wait till we are men.

With drunkenness our land is filled,
    Our homes with grief and pain;
The only free are those who from
    All poison drinks abstain.

The wise and good are praying for
   That glorious season, when
The demon drink shall be o'erthrown,—
   But, wait till we are men.

The founders of our glorious cause
   Were earnest, true, and brave,
And labored hard 'midst many foes,
   The slaves of drink to save.
Our noble leaders boldly dare
   Propose to close each den
Where drink is sold: we'll be as brave,—
   Just wait till we are men.

A noble army, brave and strong,
   Increasing every day,
Is now in training for the fight,
   Make ready—clear the way!
Boldly defying all the powers
   Of alcohol, sir, then
We'll show the world what we can do,—
   Just wait till we are men!

The Best Drinking-Place.

MARY L. WYATT.

[For either a Girl or Boy.]

ON a pleasant day in the early fall
   A stranger rode into the town,
And, stopping his horse in the public square,
   Glanced this way and that with a frown;
For the place that he sought he could not find
   (Saloons had been banished that year),
So he called to a lad who passed that way,
   And said to him, "Sonny, come here.

" Here's a nickel for you to show the way
    To the best drinking-place you know."
" All right," he answered—a quick-witted youth—
    " Just turn up that street, sir, and go
Till you come to another upon your right,
    Then turn into that and keep on
Till you come to another, turn right again,
    And you'll see it quite plainly," said John.

So, thanking the lad, the stranger rode off,
    And John gave a hop, skip, and jump;
For back came the stranger within a trice,
    Brought up—at the old town pump!
" Here you are, sir," said John with a smile;
    " The ' best drinking-place ' to be found.
Take a drink, sir; it's free, and you're welcome, too.
    It's good for your health, I'll be bound!"

He took the glass in a good-natured way
    And drank of the water clear,
Then said, " 'Tis an excellent drink, I'm sure;
    The best I've had for a year."
So saying, he tossed the lad a coin,—
    " The lesson is worth that to me.
Keep on playing your temperance joke,
    'Twill make the world better," said he.

———

The Story of Sober Dan.

REV. DAWSON BURNS.

[For a good-sized Boy.]

As through the fields I took my way,
Upon a pleasant, sunny day,
I met a little maid, and she
Was very fair and sweet to see.

Her eyes were bright as morning dew,
Upon her cheeks health's roses grew,
And from her snowy hood rolled down
Her rich long hair of golden brown.

"Good-day to you, my little maid!"
Such was the greeting that I paid,—
"Where is the home from which you came,
And what, I pray, may be your name?"
With smiling face she gave reply—
"I live in yonder house hard by;
My name, like mother's own, is Ann,
And father's now called *Sober Dan.*"

"And was he always sober, child?"
I questioned her in accents mild;
She spoke in trembling tones, and low—
"Oh, no! it was not always so;
Once father loved the evil drink,
Which brought us all to ruin's brink;
But he became a Temperance man,
And now he's known as *Sober Dan.*"

I asked, "And is home happy now?"
She answered with a cheerful brow,
"Our home is now our heaven below,
For heaven's a sober place, you know;
Father loves us, and we love him,
Now mother's eyes are never dim;
And every day to God we pray
To keep him in the Temperance way.

"And I, sir, am a Temperance lass,
And never touch the dangerous glass;
I've signed the pledge, and mother too,
And we are very staunch and true!"

"God grant," I said, "all homes may be,
Like yours, from drink and sorrow free.
And not a drinker soon be found
In town or country all around!"

A Bright Boy's Explanation.

[For either a Girl or Boy.]

A SCOTCHMAN once had a very curious dream during the night, which he told to his wife the next morning at the breakfast-table. He said he dreamed that he saw coming toward him, in order, one after the other, four rats. The first one was very fat. He was followed by two lean rats, while the fourth rat was blind. The man was greatly puzzled about his dream, and could not imagine what it meant.

He was an ignorant man and superstitious, and had always understood that to dream of rats means misfortune, and he was inclined to worry about what trouble would come upon him.

He asked his wife if she could explain the dream, but she, poor woman, could not help him. She could not tell him what the dream meant.

His son, however, a bright boy, who was busy eating his breakfast and listening to their talk, said he thought he could tell what the dream meant.

"Why, father, it's easy enough to understand that," said he. "The fat rat is the man that keeps the saloon where you go so often for beer and whiskey. The two lean rats are mother and me, and the blind one is yourself, father."

The man gave his boy a sharp look, but said nothing. He suddenly concluded that he had eaten enough breakfast, took up his hat and went off to work. I don't believe he will ask his son to interpret any more dreams for him, do you?

Margarite's Prayer.

HOPE ALTON.

[For a Girl.]

THE gloom of night was closing round
    The old house on the hill;
The icy wind went moaning by,
    The snow fell white and chill.

Within, a fire burned on the hearth,
    Its flames the only light;
The shadows flickered on the wall,
    The firelight mocked the night.

Before the fire's soft, ruddy glow,
    Close at her mother's knee,
Knelt Margarite, in gown of white,
    Her hands clasped reverently.

No curtain drawn with jealous care
    Shut in the peaceful sight,
As through the window in the door
    Shone out the wood-fire's light.

Her father, coming from the town,
    From haunts and comrades wild,
Paused, with his hand upon the latch,
    And listened to his child.

" If I should die before I wake,"
    A sharp fear seized his heart.
She looked so like an angel fair,
    "Ah, what if they must part!"

But still the low, clear voice went on,
    The simple prayer all said.
" O Jesus, look on papa dear!"
    Her father bowed his head.

"Oh, won't you bring him home just now?
    Hark! how the cold winds blow,
And mamma cries and cries so hard,
    We love poor papa so.

"And when he comes, O Jesus dear"—
    The little head dropped low—
"Don't let him talk and look so wild;
    He frightens mamma so."

A moment more she raised her head,
    A sweet look in her eyes,
"I know that Jesus heard my prayer.
    Dear mamma, don't you cry."

Her father lifted up the latch
    And came with footsteps fleet.
He knelt down at the mother's knee
    With little Margarite.

———

What Would You Think?

[For a Girl.]

WHAT would you think if the birds and the flowers
Should say that the dew and the sweet summer showers
Were not what they wanted to bathe in and drink,
They'd like something *stronger;*
Now, what would you think?

And what would you think, some pleasant spring day,
If the robin and wren and pretty bluejay
Should go reeling and falling because of strong drink
        (Just like men and boys),
Now, what would you think?

And what would you think if you picked a bouquet,
And found that the flowers acted just the same way;
And all of them tipsy because of a *drink?*
   (How queer it *would* be)
But what would *you* think?

Well, if it is silly and foolish for them,
Don't you think it is worse for the boys and the men
Who lose both their bodies and souls, too, through drink:
Now, what *do* you think?

When I am a Man.

MRS. LIZZIE DE ARMOND.

[For a small Boy.]

WHEN I am a man, I'll not worry and scold,
Or growl at the weather if too hot or cold;
I'll not use tobacco, nor drink wine or beer,
And of everything bad I'll be sure to keep clear.
I'll try for the good of others to plan,
And be a brave soldier, when I'm a man.

When I am a man, I'll let little boys
Have fun, if they do make plenty of noise;
I'll feed the beggars who stop at my door,
And give of my wealth to the ailing and poor.
I'll strive to be honest and do what I can
To make the world better, when I'm a man.

Said grandma, "Why wait till you're grown? right away
Commence your reform, begin with to-day;
You may never be old, nor rich, nor yet great,
And many a blessing you'll lose while you wait.
Strive to be and to do the best that you can,
And life will be sweeter, when you're a man.

The Cold-Water Boy.

RUFUS C. LANDON.

My friends, I stand before you a staunch cold-water boy!
Whenever thirst assails me cold water I employ.
Let others drink their cider, their ale, and wine, and beer,
I find in pure cold water a beverage to cheer!

God sends it coursing down the hills, abundant, pure, and
    free,
And what He gives indeed I think is good enough for me.
Let others sip the fiery drinks that only thirst increase,
I'll quench mine with the beverage God made for man and
    beast.

But some despise this heavenly boon, so priceless yet so free,
And quaff, instead, the poisonous draughts of sin and
    misery.
Let others sell their birthright as a thing of little worth,
I prize the pure cold water as the richest gift to earth.

———

His First Cigarette.

W. HOYLE.

" I'll be a man!" cried a giddy boy,
    As he lighted his first cheap cigarette;
He felt the flash of a new-born joy
    While the curling fumes his vision met.
Along the street with measured pace
    He strode, now gazing left, now right;
The blush upon his dimpled face
    Betrayed his pride for the new delight.

An old man stayed to see the child,
　Who scarce had left his mother's knee;
" Alas!'' cried he, in accents wild,
　"Is this child imitating me?
Come hither, boy,'' cried the grave old man,
　"Thou hast not grown to manhood yet;
Where didst thou learn such a foolish plan?
　I grieve to see this cigarette.''

The lad looked up with a knowing air,
　And shook the dust from the burning spell;
" We see gents smoking everywhere,''
　Cried he, "and like to smoke as well;
'Tis only a puff, sir, now and then,
　And when we feel a little giddy,
We think we look so much like men
　Who smoke and drink, and get unsteady.''

The old man looked at his ruddy face;
　Said he, " Thou art a young beginner;
If sin doth enter in thy case,
　I must confess that I'm the sinner.''
So saying, from its deep recess
　He pulled a meerschaum—wretched sample!
"I loved my pipe, but like it less
　For setting boys a bad example.

" Old pipe, I've told thee many a tale,
　And many a secret in my trouble,
But never thought thy charms would fail,
　Or pleasures vanish like a bubble;
Begone for ever from my sight!
　I'll smash thee in a thousand pieces!
To save the boys I'll do the right,
　For right alone pure joy increases.''

Pippin Hill.

MARY L. WYATT.

[For two characters, a large Boy and small Girl.   Fancy costumes add to the effect.]

(*Boy enters on right, walks to the front of the platform.   To audience.*)

As I was going up Pippin Hill,
  Pippin Hill was dirty;
I met a miss of summers ten;
  And I man of thirty.
  (*Girl enters on left; boy turns to her.*)
Were you that pretty little miss,
  Who dropped the courtesy sweet?
What was the paper in your hand
  You held as we did meet?

GIRL.

Oh! that was our good temperance pledge,
  And I had been to see, sir,
If any men on Pippin Hill
  Would sign this pledge for me, sir.
Of course you know that Pippin Hill
  Has many a vile saloon, sir;
I walked about till I was tired,
  From morning until noon, sir.

BOY.

Oh! yes, I know it very well,
  I've been there quite too often;
But still I think I'm not so hard
  But that my heart could soften.

But Pippin Hill's no place for you
   To show your sweet young face;
I wonder you can care for men
   Who live in such disgrace.

GIRL.

I think about their children dear,
   And poor, heart-broken wives, sir,
And think if they would sign the pledge
   They'd lead far better lives, sir.
    (*Presents paper and pencil to him.*)
Now, won't you kindly write your name
   Upon that line to-day, sir:
And promise not to smoke nor drink,
   And keep the pledge always, sir?

BOY.

My little miss, my pretty miss,
   Blessings light upon you;
If my pockets were full of gold
   I'd spend it all upon you.
    (*Takes paper and pencil.*)
Yes, I will sign your temperance pledge,
   You ask so very sweetly;
And never drink, nor smoke, nor swear,
   But be reformed completely.

GIRL.

Oh! you are very kind, indeed,
   To give so prompt an answer;
I think you're very good, I'm sure,
   And quite a noble man, sir.
Now, won't you try and help us make
   Some prohibition law, sir?
Don't spend your money, sir, on me,
   But spend it on the cause, sir.

The First Drink.

GEO. W. BAIN.

[For a large Boy.]

WE are told "there is no moral wrong in a glass of beer *per-se*."

What is *per-se?*

Per-se is, *by itself.*

A horse loose in a pasture is *per-se*. When you bridle, saddle, and mount him he relates to something, and is not *per-se*.

There is no moral wrong in a glass of beer *per-se*. That is, there is no moral wrong in a glass of beer by itself. But when a man swallows the beer it is not *per-se* any longer.

Then it goes to join the procession of drinks and becomes accessory to the sin of drunkenness.

I would like to show the boys the relation of the first drink to the sin of drunkenness.

You will agree with me when I say drunkenness is a sin.

Now if it takes six drinks of liquor to make a man drunk, where does the sin come in? Is the first, second, third, fourth, fifth, innocent, and the sixth alone guilty? Does the sixth bear all the sin ?

Some one says, no, it begins with four; another says three.

Suppose an incline plane running from this platform to a fearful precipice, over which, if a boy falls, he is dashed to death.

Suppose I stand at this end of the incline, a man at the other, and four between at equal distances representing the six drinks.

Suppose I say to a boy, "Come here and take a slide," and you know the temptation to coast is great with a boy.

The boy says, " Isn't there danger down yonder, at the other end ? "

" Yes ! but get off before you get there," and I give him a shove.   He reaches the second ; he gives another shove and he goes faster.

When he reaches the third he says : "Sir, is there danger ahead ? "

"Yes ! but get off before you get there," and another shove sends the boy on his downward course.

The fourth increases the speed by another shove.

When he nears the fifth he cries out : "Sir, is that a precipice yonder ? "

" Yes, but get off if you can," says the fifth, as he sends him on with another shove.   In an instant the sixth seizes the boy and hurls him over to his death.

Is the sixth man a murderer and the others innocent ?

Where did the danger begin ?

With the first shove.

So on the incline plane to the sin of drunkenness the sin commences with the first drink.   " Abstain from every appearance of evil "; to drink is not to abstain.

Boys, let your motto be, " Moderation in regard to things useful and right, total abstinence in regard to things hurtful and wrong.

Malty and Isadore.

A STORY WITH A MORAL.

MRS. VIRGINIA J. KENT.

[For either a Girl or Boy.]

Two favorite cats, being cold one day,
The oven door open, and mistress away,
Stepped softly in and laid down for a nap,
As serenely as if in their mistress' lap.

How long they slept, they never knew
(Nor any one else, if the story be true),
For while they quietly purred, and dozed,
Some one entered and suddenly closed
The door; and of their terrible fate
No one knew, until *too late*.

"Where are Malty and Isadore?"
The mistress said, for never before
Had they strayed away from their home like this—
"There surely must be something amiss."
And so there was; as she went to inquire,
Bridget made up a great big fire;
The kettle sang and the tea was made,
And everything for the supper laid,
When from the oven in the range
There came an odor so very strange;
The door was opened.   Oh! such a sight!
It filled the family with affright.

There were Malty and Isadore,
Two burned coals—and nothing more.
They had just stepped in for heat and repose,
And what they suffered no mortal knows.

   .   .   .   .   ;

There are places more dreadful on every hand
Than ovens thrice heated throughout our bright land.
They are drawing and coaxing our youth in their snare;
Let all the dear children avoid them with care.
There are places which dazzle, but lead unto death,
And are cursing our loved ones with poisonous breath.
Be watchful, be wary, look not on the wine,
For it dances and sparkles with treacherous shine;
There is death, and destruction, and undying woe
In the place where the *Bible says* all drunkards go.

A Sensible Dog.

CORA B. TAYLOR.

[The Boy who recites this should have a dog near him, which may be either lying on a chair or held by a string.]

Now here's a dog whose head is clear,
He drinks no wine or lager beer;
I'll warrant that he'll have more knowledge
Than many drinking men in college.

Now, friends, don't think I'm here to joke,
But this same dog refused to smoke;
Tobacco, too, he does not chew;
I think he's sensible, don't you?

And if a dog of brilliant mind
To drink and smoke is not inclined,
I think 'twere better you and I
Should pass such evil habits by.

So, sir, if you will come with me,
And one among our number be,
I think you'll find there is no need
To drink, or even use the weed.

Which Road Will You Take?

ANNA LINDEN.

[For a good-sized Girl.]

Come, boys, take the floor! stand up manly and true,
And tell me just what you are going to do;
The world seldom notices how you are dressed—
If it sees you are clean it excuses the rest;
But it gives careful heed to your actions and talk,
To discover the path you are going to walk.

Two roads lie before you.　The one on the right
Has honor and manhood all plainly in sight;
The other grows darker each step that you take,
Till sin, crime, and punishment make the heart ache.
You would not like that, with the sin and disgrace
To sadden your heart and disfigure your face.

What are brains made for?　"For use," you will say,
"For thinking, and storing up knowledge each day."
And what are hands made for?　Your bright looks reply:
"To do what we can and do more by and by."
Your answers are good, so keep the right track,
Onward and upward, and do not turn back.

What are mouths made for?　You surely can tell,
And answer the question both quickly and well;
They are made for clean words, clean food, and clean drink,
And made for bright smiles—isn't that what you think?
But they never were made for a gateway of sin,
To pour liquid fire and poison within.

Your lips should be sweet for a good loving kiss
For mother, or auntie, or dear little sis;
So don't use tobacco, for often it leads
To the road on the left, with bad boys and bad deeds,
And makes people think, as a general rule,
Such a boy needs no dunce-cap to show he's a fool.

———

A Bit of Common Sense.

THE life of man is but a span,
　　And whiskey makes it shorter;
But stores of wealth and rosy health,
　　Bless him who drinks cold water.

Walk Steady.

MRS. M. A. KIDDER.

[For a good-sized Girl.]

I saw a poor man,
  Who had only one leg
And one hand—so, you see,
  He did nothing but beg;
And wanting his crutches,
  He stirred not a peg!

Yet I know a worse case,
  Quite sad to relate,
Of a man who had two legs,
  Yet couldn't walk straight,
But staggered and reeled—
  A most terrible state!

He'd start off all right.
  Yet frequently pause
To drink by the way;
  So rum was the cause
Of his shaming his dear ones,
  And breaking the laws.

Now, dear boys, look out,
  You've sturdy young feet;
Stand straight, and be manly,
  In highway or street!
Yield not to temptation,
  And keep your breath swect.

"Can You Undo?"

MIRIAM RUTH SMITH.

[For a good-sized Girl.   This should be spoken with care, the speaker
pathetic and earnest.]

IN a hospital bed a soldier lay
Wounded and sick at the close of day,
Moaning and tossing in feverish pain,
Trying in vain some ease to gain.

His heavy groans attract the ear
Of one who is trying the sick to cheer
With the old, old story of Jesus' love,
Leading their thoughts to a rest above.

" My friend," he said in accents low,
" Is there anything I can do for you ?
Is the pain so bad ?   Is the soul unblest ?
What can I do, lad, to give you rest ? "

" Do !" said the soldier, excitedly,
" Do for me, sir ! did I hear you say ?
I'm sure it's very kind of you,
But tell me this, sir, can you undo ?

" Last year to our regiment a young lad came,
With a pure young heart, and a spotless name ;
Each night he knelt by his bed in prayer,
And this was more than my spirit could bear.

" I told him I'd have no canting there,   -
And then I began to curse and swear ;
I vowed, sir, I'd make that young lad's life
With every kind of evil rife.

" My influence, you see, sir, was very strong
  Over a lad so timid and young ;
  I soon laughed him out of the praying line,
  And got him to follow more in mine.

" I introduced him to what I called life—
  Drinking, smoking, gambling, and strife ;
  In one short year you could hardly tell
  The lad who had been trained so well.

" We fought in the battle yesterday—
  Stood side by side in the deadly fray ;
  The bullets flew thickly, far and wide,
  And that lad was struck down at my side.

" He fell, sir ; but where is his soul, oh ! where ?
  Has he gone to the depths of dark despair ?
  It was I that made him that path pursue—
  Oh ! tell me, sir, tell me, can you undo ?

" The loss of his soul will be charged to me—
  My influence led him wrong, you see.
  You cannot help, sir ; it's kind of you
  To trouble so much, but you can't undo."

Oh friends ! what a lesson for us to learn !
Are we sure our influence will never turn
A soul astray from the path of heaven,
To die at our side with sin unforgiven ?

Life's road is beset with many a thorn :
Let us make it our joy to keep some down ;
Let us brighten the path that we daily pursue
By an influence right ; for we cannot undo.

Broken Jugs.

[For a bright Boy, who should step on the platform laughing.]

HA! ha! ha! I can't help laughing when I think of something that Cousin Ben just told me. I'll tell it to you, so you can laugh too.

"A man named Joe sent for a jug of rum. Rum was not sold in his town. He sent for the jug to come by the cars. As he took the jug from the cars he let it drop on the track, so jug and rum were lost. Then he sent more money, and got a new jug of rum. On his way home with this jug in his hand he fell on the ice. He fell on the jug. The jug broke in two as it hit the walk. Then all that rum was lost also. Then he sent for a third jug of rum. Some men on the cars knew that when Joe had no rum he was a quiet man. But when he had rum he was cross and bad to his wife. The men thought they would play a trick on Joe for his good. So they threw out all the rum and filled the jug with milk. When Joe got home he found that he had a jug of sweet milk. His wife was glad of that. Well, he said he would try again; so once more he sent for a jug of rum. When he got it he set out for home. His home was two miles from the cars. As he went home with his jug he felt as if he must have a drink at once. He was very cold. He set the jug by the side of the road. Then he went over a field to a house. He told the girl at the house that if she gave him a mug of hot water and some sugar, he would come next day and cut wood for her. The girl gave him the mug, the water, and the sugar. He ran back to the road where he had left the jug. But—the jug was gone! He has not seen it since. Now he keeps sober. He says he is tired of trying to get jugs of rum."—*Adapted from "Temperance First Reader."*

Katie's Thanksgiving Letter.

ALICE MAY DOUGLAS.

A LETTER once poor Katie wrote,
  And on its way it sped
One bright Thanksgiving morning.
  'Twas thus the letter read :
" Oh, farmer man ! Oh, farmer man !
  Do please to come this way,
Because we want a turkey
  On this Thanksgiving Day.
Oh, do you think that none of us
  Here in this narrow lane
Have nothing to be thankful for,
  In spite of toil and pain ?
I have two hands with which to work,
  Two feet with which to walk ;
And I can hear, and I can speak,
  And with my mamma talk.
And when I'm cold and hungry,   -
  I then can sing a song
And think I'm warm.   When headaches come
  They never do last long.
With so much to be thankful for,
  I'd keep Thanksgiving Day ;
So bring a turkey, and sometime
  You'll surely get your pay.
Leave it at Bragg's Lane, number five,
  And please wait for my thanks."
The postman gave this letter
  To crabbed Farmer Hanks ;
Who hung his biggest turkey
  That day on Katie's door.
With it this note : " You've made me, child,
  More thankful than before."

He Knew What It Meant.

JESSIE FORSYTH.

In a quaint old church in a western town,
On a pleasant eve as the sun went down,
Some boys and girls and some women and men
Heard the pitiful story told again

By a lady who spoke about the drink ;
Of the ills it wrought, and she bade them think
Of some work to do that would help along
The righteous cause and defeat the wrong.

She said that the hope for the future lay
In the boys and girls of the future day.
And she begged that, to arm them for the fight,
They would join the temperance band that night.

Then a good old man rose and said, " I doubt
If these children know what it's all about.
You had better wait till they understand
'Bout a pledge for life ere they join a band."

But a boy of seven, brave and ready,
Said with eager voice, but purpose steady :
" I know what it means ; I mean to do it :
It is saying a thing, and *sticking to it !* "

Which Will You Drink ?

Which is most beautiful, pure and good,
    Wine drops or water drops ?   Which, do you think ?
The very best drink for a thirsty world is water ; not wine
        drops.
    Which will you drink ?

"A Suit for a Song."

ELEANOR W. F. BATES.

[This is best adapted for a Girl of good size who can render it with pathos.]

'TWAS a cold winter's morning. The great clothing store
Had folded its shutters and opened its door.
The full ranks of salesmen were busy as bees,
For patrons were many to fit and to please.
The large plate-glass windows were shining and bright,
And behind them arranged was a wonderful sight—
Piles of clothing galore, both for boys and for men,
While mirrors each side showed their glories again ;
And a great gilded sign (broad its letters and long)
Bore this legend enticing, " A Suit for a Song."

The master of all, the rich merchant, stood by,
Prosperity shown by his keen business eye,
His carriage erect and imperative hand,
As he glanced right and left with an air of command.
While he stood, through the door crept a mite of a boy,
Not one of the dainty curled darlings of joy,
But a ragged and dirty and half-frozen child
Looked up at the merchant and timidly smiled ;
And then, like a charm of far bells set a-swing,
Half-murmured, half-whispered, " Please, sir, may I sing ? "

He sang, and his voice trembled sweet on the ear ;
He sang—oh, the angels might bend down to hear !
'Twas the lyric of childhood, and passionate pain
And joy's magic music were mixed in the strain.
It was low—'twas the cry of a heart-stricken sore ;
It was soft, and the ardor of faith went before ;
It was shrill ; tears unbidden sprang swift to the eye,
For cold and starvation rang keen in the cry ;

It was sad with the pleadings of hope long deferred,
Yet 'twas sweet as the lay of a nest-building bird ;
Yes, 'twas sweet ; it flung memories of home on the air,
Of purity's shrine, of a mother's low prayer ;
It faltered and failed into silence, and then,
Looking round at the circle of listening men,
He said—though his voice for a moment fell mute—
" I've sung you a song—will you give me a suit ? "

He pointed his thin, grimy finger to where
The sign in the window was lustrous and fair.
" A Suit for a Song "—it was this the child meant ;
Every eye on the prosperous master was bent.
He spoke not, he moved not.   Far back in the years
He roamed with a vision sweet almost to tears.
His face was downcast on the quivering child,
But in one moment more he had looked up and smiled,
And patted the boy.   " I suppose I'm a fool—
Here you ! dress this imp in a suit fit for school,
And the rest of you fellows " (with mimic berating)
" To your work ! and be quick ; there are customers wait-
        ing ! "

All day was the heart of the merchant prince warm
As the suit that now covered the little one's form ;
And whenever the issues of business perplexed
His brain to confusion, a wandering text
From an old-fashioned volume brought peace out of strife,
And calm and content to an oft-worried life ;
" Naked I was, and ye clothed Me "; the words
Chorded sweet as a chorus of jubilant birds—
Nay, sweeter ! as faith is far sweeter than joy,
They were sweet as the song of that newly-clad boy.

How it Pays.

MARY E. BRADLEY.

SAID Tom to Dick and Harry, "The wind is sharp to-day;
Suppose we have a whiskey straight, to keep the cold
away?"
"All right"—the cheerful answer—"that's just the talk
for me!"
And the smiling landlord mixed the drinks and pocketed
the fee.

Another day the comrades met at the door again;
And now 'twas heat instead of cold that made them all
complain:
"Thermometer at ninety and such a blazing sun!
Let's have a drink to cool us off":—no sooner said than
done.

There stood the smiling landlord—in his button-hole a
flower;
He mixed for Tom a "whiskey straight," for Dick a
"whiskey sour";
And when he found that Harry preferred a "brandy
smash,"
He mixed it with as good a grace—and pocketed the cash.

A boy looked on and wondered (a boy that was no fool),
How drink could warm men up one day and one day make
them cool.
"It doesn't stand to reason, the thing can't work both
ways."
The smiling landlord answered him, "No matter, if it pays.

"The whole thing's in a nutshell—when people want to
drink
It warms them up or cools them off, just as they choose to
think.

It pays; that's all I care for." The boy thought, " Yes,
 that's so;
But how it pays the other folk is what I want to know."

'Twas easy to discover, for the downward road is quick
To men who drink for heat and cold like Harry, Tom, and
 Dick.
Their business went to ruin and they to want and shame;
But the landlord mixed his liquors and sold them all the
 same.

And so the boy learned wisdom, " He shan't grow rich on
 me;
For I'll quench my thirst with water, God's own free gift!"
 thought he.
He kept his word, and prospered in honest, sober ways;
And, rich in health and happiness, his life shows how it
 pays.

———

What Shall We Talk About?

W. A. EATON.

[This can be recited by a Girl or Boy, or divided among four, each
reciting two verses.]

WE were sitting by the fire
 One dreary winter night,
It was so warm and cosy
 We would not have a light;
Mamma was singing softly
 About the " better land,"
While baby Jim was counting
 The fingers of his hand.

The timepiece on the mantel
   Had chimed the hour of six ;
Mamma said, " Bring the kettle,
   And two or three dry sticks—
For Papa will be coming,
   And he will want his tea " ;
And soon the well-dried fagots
   Were blazing merrily.

When Papa came, he shouted,
   To find us all so still.
" I thought I'd got the wrong key—
   Is anybody ill ?
Is it a Quakers' meeting—   •
   Why are you all so quiet ?
My girls, why don't you chatter ?
   Now, boys, get up a riot ! "

But Nelly said, demurely,
   " Oh, Papa, girls don't shout ;
The fact is we were waiting
   For something to talk about."
Then Papa said, " Well, really
   You fill me with surprise—
Why here are many subjects
   Laid just beneath your eyes.

" The coal within the firegrate—
   What matters here for talk :
The world's great age stamped on sandstone,
   On coalpit, and on chalk.
Of firedamp and explosion,
   Of dreadful loss of life ;
Of hours of anxious waiting
   For the collier's child and wife.

"The kettle lid now dancing
    Under the power of steam,
Tells of Watts' great invention,
    We hear an engine scream.
The hearthrug will remind us
    Of Arkwright's patient life,
How all his great inventions
    Were laughed at by his wife.

"The loaf upon the table
    Suggests the waving grain,
The anxious farmer watching
    For the first sign of rain.
The tea and sugar also
    Suggest China or Assam ;
And, oh ! what food for thought lies hid
    In one small pot of jam."

So, girls and boys, remember,
    Wherever you may be,
Some lessons you may gather
    From everything you see :
And when you see a drunkard
    Tottering on ruin's brink,
Bid all who come within your reach
    Never to touch strong drink.

An Eloquent Description of Water.

[For a bright wide-awake Boy, who should speak with decision and energy.]

YEARS ago Paul Denton, a Methodist preacher in Texas, advertised an out-door meeting, with better liquors than are usually furnished. When the people in the crowd were assembled, a desperado in the crowd cried out, " Mr. Paul

Denton, your reverence has lied! You promised not only a camp meeting, but better liquor.   Where's the liquor?"

"There!" answered the missionary, in tones of thunder; and pointing his motionless finger at the matchless double spring gushing up in two strong columns, with a sound like a shout of joy from the bosom of the earth, "There," he repeated, with a look terrible as lightning, while his enemy actually trembled before him.   "There is the liquor, which God, the eternal, brews for all His children.

"Not in the simmering still, over smoky fires, choked with poisonous gases, surrounded with the stench of sickening odors and corruptions, doth your Father in heaven prepare the precious essence of life, the pure, cold water; but in the green glade and grassy dell, where the red deer wanders, and the child loves to play, there God brews it; and down, low down in the deepest valleys, where the fountain murmurs and the rills sing; and high up on the mountain tops, where the naked granite glitters like gold in the sun, where the storm-cloud broods, and the thunderstorms crash; and away far out on the wild sea, where the hurricane howls music, and the big wave rolls the chorus, sweeping the march of God,—there He brews it, that beverage of life, health-giving water.

"And everywhere it is a thing of life and beauty, gleaming in the dew-drop; singing in the summer rain; shining in the ice-gem, till the trees seem turned into living jewels, spreading a golden veil over the setting sun, or a white gauze around the midnight moon; sporting in the cataract; dancing in the hail shower; sleeping in the glacier; folding its bright snow curtains softly about the wintry world, and weaving the many-colored sky, that seraph's zone of the sky, whose warp is the rain-drops of earth, whose woof is the sunbeam of heaven, all checked over with celestial flowers by the mystic hand of refraction.   Still always it is beautiful, that blessed life-water! no poisonous bubbles

on its brink; its foam brings not madness and murder;
no blood stains its liquid glass; pale widows and starving
children weep not burning tears in its depths; no drunkard's shrinking ghost from the grave curses us in worlds of
eternal despair! Speak out, my friends; would you ex-
change it for the demon's drink, Alcohol?"

A shout, like the roar of a tempest answered, "No!"

––––––

Praying for Shoes.

A BOY'S THANKSGIVING.

PAUL HAMILTON HAYNE.

[For either a Girl or Boy.]

ON a dark November morning
  A lady walked slowly down
The thronged, tumultuous thoroughfare
  Of an ancient seaport town.

Of a winning and gracious beauty,
  The peace on her pure young face
Was soft as the gleam of an angel's dream
  In the calms of a heavenly place.

Her eyes were fountains of pity,
  And the sensitive mouth expressed
A longing to set the kind thoughts free
  In music that filled her breast.

She met, by a bright shop-window,
  An urchin, timid and thin,
Who, with limbs that shook and a yearning look,
  Was mistily glancing in

At the rows and varied clusters
  Of slippers and shoes outspread,
Some, shimmering keen, but of sombre sheen,
  Some, purple and green and red.

His pale lips moved and murmured;
  But of what, she could not hear.
And oft on his folded hands would fall
  The round of a bitter tear

"What troubles you, child?" she asked him,
  In a voice like the May-wind sweet.
He turned, and while pointing dolefully
  To his naked and bleeding feet,

I was praying for shoes," he answered;
  "Just look at the splendid show!
I was praying to God for a single pair,
  The sharp stones hurt me so!"

She led him, in museful silence,
  At once through the open door,
And his hope grew bright, like a fairy light
  That flickered and danced before.

And there he was washed and tended,
  And his small, brown feet were shod;
And he pondered there on his childish prayer
  And the marvelous answer of God.

Above them his keen gaze wandered,
  How strangely from shop and shelf,
Till it almost seemed that he fondly dreamed
  Of looking on God himself.

The lady bent over, and whispered :
" Are you happier now, my lad ? "
He started, and all his soul flashed forth
In a gratitude swift and glad.

" Happy ?—oh, yes !—I am happy ! "
Then (wonder with reverence rife,
His eyes aglow and his voice sunk low),
"Please tell me !   Are you God's wife ?"

————

A Plea for a Collection.

[For either a Girl or Boy.]

BEFORE our meeting closes, allow me just a word.
We hope you've been amused and pleased with all that you
    have heard ;
But now it rests with you to crown our efforts with per-
    fection,
Please show your sympathy to-night by a right good col-
    lection.

Don't say we're always begging—it really isn't true !
We can't manage without money more than other people
    do.
It's for us boys and girls, and you will gain our best affection,
If you will favor us to-night with a tip-top collection.

We thank you very much indeed for coming here to-night,
And hope we've entertained you—we've tried with all our
    might ;
But if in what we've said or done, you see some slight
    defection,
Still, overlook our faults, and give us a generous collection.

It's nice to hear you clap and cheer, and laugh at what is
    funny,
But now we want to hear as well the chinking of your
    money !
Don't disappoint our hopes, dear friends, and plunge us in
    dejection,—
But hear our plea and make us glad by a superb collection

———

What Came from Signing the Pledge.

VIRGINIA J. KENT.

[For either a Girl or Boy.]

WE have lots of nice things at our house these times,
    And lots of money, too ;
So many nickels and so many dimes,
    We hardly know what to do.

For papa has changed his mind, you know,
    And instead of throwing away
A nickel here and a nickel there,
    A dozen times a day,
In those horrid saloons, as he used to do,
    He brings them every one home ;
Dear mamma has the most of them,
    But Susie and I have *some*.

We often used to be hungry and cold,
    And mamma would cry *all* day,
And papa used to be naughty and cross,
    And we had to hide away ;
But now he's as good as good can be,
    And seems to love us so ;
He don't leave us alone at night any more,
    As he used to long ago ;

He says it's all from signing the pledge—
　　　What a splendid thing it must be!
To bring again such happiness
　　　To mamma, and Susie, and me!

The Bottle Tree.

LUELLA DOWD SMITH.

[For either a Girl or Boy.]

In Australia grows a strange tree with an enlarged trunk. Because of its shape, it is called the "Bottle Tree."

In our own land grows the deadly Alcoholic Bottle Tree. Its breath is more poisonous than the Upas Tree. Its shadow is more noxious than the deadly nightshade.

"Oh, evil was the root; and bitter is the fruit."

In the great book of nature God has written concerning this tree—"In the day that thou eatest thereof, thou shalt surely die." All through the ages mocking voices have protested—"Thou shalt not surely die." Men have yielded to the deceit, gathered the deadly fruits of alcohol and stumbled down into their graves. Thus in our land eighty thousand or more every year go down to death. Two million five hundred thousand are staggering under the curse of this deceitful tree; poor, wretched, blind; slaying the friends who love them most, drinking the blood of hearts and destroying souls. Ten million in our land to-day are sorrowing in the shadow of the American Bottle Tree.

Its poison makes idiots and lunatics, paupers and criminals. Does some one ask, "Why does not our nation root up, and forever destroy these Bottle Trees?" You do not know, then, the selfishness of this nation and the greed of this age.

Our nation protects the trees.   It has a royalty from all their fruit.

"Our nation deals in the blood of her children!"   Did some one say that?   Perhaps conscience said it.   Perhaps God said it.

But our nation does not listen.   Men are cultivating the Bottle Tree.   Its fruit grows stronger by the care, stronger and more deadly.

The nation's children eat of it and die.   Nay, more; our nation sends the fatal fruit abroad into the dark corners of the earth, to poison the heathen who sit in darkness and ignorantly reach out their hands to take and eat and die. And the money received for the poison fruit is the price of blood.

Oh, friends, we appeal to you.   See the evil of this deadly tree, and for the love of humanity, for the love of Christ, resolve you will refuse its envenomed fruit, and try to avoid its poisonous breath and blighting shades; but more than this—that you will, with all the power God gives you, smite this monstrous growth, tear up its crawling roots, strike down its baneful trunk, and thus do all you can to save our beloved land from its deadliest scourge, the poison Bottle Tree!

———

When I am a Boy.

MRS. E. A. HAWKINS.

If, when I'm a boy,
I am lazy, and shirk
My work upon some one that's smaller,
The chances are good
I shall do the same thing
When I have grown older and taller.

If, when I'm a boy,
I am always behind,
And never make any advances,
When I am a man,
Some one else, and not I,
Will be sure to get all the best chances.

If I use, when a boy,
Cigarettes, and talk slang,
Without either thinking or caring,
You will probably find me,
When I am a man,
Chewing navy tobacco and swearing.

If, when I'm a boy,
I drink cider and beer,
And persist, against reason and warning,
You may find me in rags,
And as drunk as a sot,
Fast asleep in the gutter some morning.

Now, that's not the *kind*
Of a man *I* would make;
The world has too many already:
So I will begin,
Right away, while a boy,
To be temperate, honest, and steady.

Oliver's Dream.

MRS. J. MCNAIR WRIGHT.

A YOUNG man named Oliver was a stonemason. He had built a large wall for a liquor-seller. The man called Oliver into the bar-room to get his pay. "You had better drink good luck to the wall," he said to Oliver.

He had never taken any strong drink, but Oliver had not courage to say "No," so he drank a glass of whiskey and treated all the men in the bar-room.

After that he started for home. He felt sick; his eyes burned, his head was dizzy, he could not walk straight. He knew quite well that he was not sober, and he was truly ashamed of himself. When he got home he went immediately to bed.

When Oliver's little sister came to call him to tea, he said, "Go away, Ann, my head aches; I do not want any supper." Then he lay thinking of all that he had heard at a temperance meeting, the night before; and so he fell asleep. He had a very strange dream.

Oliver dreamed that all the different parts of his body began to talk, and he lay and listened to them. First his mouth cried out: "What was that horrid stuff Oliver poured into me? I never tasted such a thing before! It burned my tongue and my throat, and left a bad taste, and I hurried it down as fast as I could, to get rid of it!"

" I should say you did," growled Oliver's stomach. "You hurried it into me, the worse luck! Why, I am fairly blistered! I am burning with thirst, and am covered with red blotches. I'm sick. My work is to digest food respectably, but that stuff stopped my work, just as you might stop the turning of a wheel, by driving a spike into it. If I cannot get rid of that stuff, I shall be ruined."

"Since you know how bad it is," shouted Oliver's head, "why were you so mean as to send it to me? You know how very important I am. How can the arms or legs move without my orders? Here you have sent me that liquid fire, and it has about ruined my delicate texture. I am very sensitive, let me tell you, and I cannot stand such rough treatment. My eyes are burning; my ears are roaring; I am dizzy; I feel like a wheel of fire. I cannot send out proper orders to the rest of the body."

"I should say you could not," cried Oliver's legs in wrath. "Here we feel as big as trees, and as heavy as lead! How did we come home? Why, we took up all the side-walk, flopping about, as if we were half paralyzed! It is a wonder we did not fall down and get broken! Next time we are treated so badly, we may give out, and who knows, Mr. Head, but you will be cracked on a curbstone?"

"You are no worse off than we are," moaned Oliver's hands; "all our fingers feel as big as cucumbers. We could not unbutton Oliver's vest; we dropped his cravat; we pulled his shoe-strings into knots; we could not hang up his coat; there it lies on the floor. We ache, we are stiff; oh, dear!"

"I have as much right to complain as any of you," wailed Oliver's blood; "can't you see I am poisoned? I am hot, I am thick, I am dull, I cannot flow along properly; and soon I shall show how miserable I am by sending a lot of spots, and sores, out on the skin, and then you will all be angry at me. But it is not my fault."

"It is the fault of that dreadful burning drink," thumped Oliver's heart; "I felt it in a minute. Here all day I have worked and pumped, and fairly stood the strain, while Oliver lifted stones and climbed walls. I expected to get a let-up when evening came. I need rest; I was looking for-ward to it; but here I am working harder than ever! That drink sends me flying! If this goes on, my walls will grow thin, or break. Or perhaps this hot, bad blood will bring on some other disease, and I shall stop working. Then where will the rest of you be?"

"Don't leave me out of the count," said a little squeaky voice; "I am Oliver's purse, and I want to be heard. I am very thin to-night. I ought to be full. I expected to go to the bank and leave a nice roll of bills there to-morrow. I had been told I was to go to the store, and buy Ann a dress. Oliver spent more than you think on that whiskey, and then, with what he had drunk, he got stupid, and the man did not give him right change by two dollars, and cheated him out of a five-dollar bill in the pay! And Oliver signed the receipt, and here it is, sticking in my throat, like a fish-bone, when I know I never got the money!"

After this dream Oliver woke. It was night, and the moon shone. He went softly down stairs, put his head under the hydrant, and let water run on it for a long time. Then he drank a quart of cold water, and carried a pail of water to his room, and had a bath. Then he got on his knees, and vowed that he would never touch another drop of strong drink.—From " *Temperance Third Reader.*"

Our Jetty.

MRS. M. A. KIDDER.

[For a Boy.]

OUR cat is as black as a coal, if not blacker,
And she'll eat many things, from a mouse to a cracker;
She'll eat a sweet cake or a bun from the baker,
But there's one thing she won't eat, and no one can make
    her.

She'll beg for mince pie, or pudding when handy,
But she won't touch a morsel that's seasoned with brandy.
I say to mamma, and quite often to Betty,
"Was there ever a cat like our own darling Jetty?"

I'm a temperance boy, with a home in the city,
And I'm one, heart and soul, with our teetotal Kitty.

Tom Linton.

TOM LINTON was a temperance boy,
  With heart so blithe and free ;
He got a holiday one day
  And climbed into a tree.
He listened to the birds on wing
  And plucked the blossoms gay,
While in a cheerful voice he sung
  This merry temperance lay :
" The little birds that fly and sing,
  So happy, blithe, and free,
Are water drinkers, every one,
  Teetotalers true like me.

" The lovely flowers that bloom so bright
  In hues so rich and rare,
Drink only water from the skies,
  And I their drink will share.
The trees that grow so tall and strong,
  And spread their branches wide,
All quench their thirst from dews and showers,
  They, too, are on our side.
Yes, birds and flowers, and stately trees,
  And beasts that walk the sod—
All nature's with us, and our cause
  Is blessed by nature's God."

Who'll Kill King Alcohol?

" WHO'LL kill King Alcohol ? "
  " I," says Prohibition,
  " I'm in just the position,
    And if I'm backed by the Constitution
I'll kill King Alcohol."

" Who'll see him die ? "
  Says every temperance man, " I !
  And we won't heave a sigh ;
  We'll be glad to see him die."

" Who'll catch his blood ? "
" There would be such a flood ;
  Let it fiow into the mud.
  For it's not any good ;
  So we won't catch his blood."

" Who'll make his shroud ? "
  List to the voices loud :
    " All over the land
    We'll take a hand
  To help make the shroud."

" Who'll toll the bell ? "
" Oh, we won't toll a bell,
    But we'll shout loud and clear,
    So that every one can hear ;
  And 'twill ring through the dell,
  But we won't toll a bell."

" Who'll dig his grave ? "
    " I," says the drunkard.
" I've always been his slave ;
    I'll dig it long and deep,
    May he forever sleep !
  I'll be glad to dig his grave."

" Who'll be the chief mourner ? "
    " He has made so many cry,
    Now he's about to die,

There's not one far or near,
That could shed a tear,
Or weep o'er his bier;
So there won't be a mourner."

What Can We Do?

MRS. VIRGINIA J. KENT.

[For four little Girls.]

FIRST CHILD.

I KNOW of a home, a desolate home,
Where joy and happiness seldom come,
Where the chairs are broken and window panes out,
And things are thrown in disorder about;
Where the children look dirty and poorly clad,
Where the mother looks tired, and wan and sad,
And the father—oh, that is the trouble, you see,
And this is the secret between you and me,
He spends all his time, and his money too,
In a place where good people never do,

SECOND CHILD.

Why, how strange that is! and where can it be
That he goes?—to cause such misery.
He must be a very wicked papa
If he knows how unhappy his children are,
That he doesn't stay with them, and give them food,
A pleasant home, and everything good;
I thought that was what all papas were for;
I wish he would never do so any more.

THIRD CHILD.

I wish we could shut up that dreadful place,
That causes such misery and disgrace;
But what could little girls like us four
Do toward closing a grog-shop door?

It is true there are some things that children can do
To help wicked men to be good and true ;
One is to coax them to church to go,
Another, to sign the pledge, you know ;
I have heard of men that were reckless and wild
Being led by the hand of a little child
Away from sin into Jesus' fold,
The only safe place for young or old.

FOURTH CHILD.

I have just a word to say ;
All we little ones can pray
To God, who rules above the sky,
And hears His children, when they cry.
He pities the mothers whose hearts are breaking,
And the little children whose hearts are aching.
He only can remove the cause,
And make the people keep His laws ;
Then let us go to Him—just now,
And at His feet with reverence bow.

(Children with bowed heads repeat in concert.)

Dear Father, from Thy mercy seat,
Look down, as humbly at Thy feet
We little ones are bowed in prayer ;
O wilt Thou in Thy mercy spare
Those who are now upon the brink
Of ruin, from this fearful drink ?
Shut the saloons in all our land
And raise a mighty temperance band ;
Keep all our papas from ever taking
The wicked stuff, and stop the making
Of poisonous drinks, that people take.
We ask it all, for Jesus' sake.

                    Amen.

Speech for Opening of Meeting.

[For either a Girl or Boy.]

DEAR FRIENDS: In behalf of this Society, I bid you welcome. We are happy to have your presence, and we hope to merit your approval and receive your encouragement. We are a small band of this great temperance army enlisted for life. Our banner bears the inscription: *No Intoxicating Drinks, No Tobacco, and No Profanity.* We hope to learn the history and principles of the temperance reform so fully, and practice the rule of temperance so faithfully, that there may not be one drunkard among us. We not only wish to become better ourselves but to help others. A motto for each member is, "*I'll try to bring one.*" Our members will try to entertain you to-night with speech and song. We know that you will remember we are only beginners, and look with charity upon our efforts We are young in years, but our hearts are in the work, and our determination firm to help on the temperance cause, first by being right ourselves and endeavoring to get others to promise to touch not, taste not, handle not, any kind of strong drink.

Minnie is Sure to Know.

T. H. EVANS.

[For a little Girl.]

I'M just the oddest, merriest romp that ever you did see;
At least that's what my mother dear is always telling me.

I can't tell why I'm such a tease, and have such tiresome
    ways:
I only know it can't be helped, I've been so all my days.

They say I'm just no good at all, except to laugh and play;
I'm fond of fun, and very small, and always in the way.

But one thing I should like to say, that you've not heard
before ;
I'm not so small but what I make just one teetotaler more.

I've heard such heaps about the drink and all the harm it's
done,
When I get near a drink saloon, you ought to see me run !

There's something people say is " Woe," I've heard from
Uncle John ;
I don't know 'zackly what it means, but drinking brings
it on.

There's something, too, called " Poverty "—no food nor
clothes at all ;
At least I think that's what it means—and drinking does
it all.

And " Crime " I 'member's something else, of which I've
often heard ;
And " Madness " too, I fancy, is another drinking word.

I think you'll own now, after that, which is not all by far,
Such drinks are bad for little girls—now, don't you think
they are ?

And if that's true, they're bad for you ; they must be, as a
rule ;
You're only boys and girls, you know, who've grown too
big for school.

So come and sign the pledge to-night, before you leave
the hall ;
For if you don't, I might as well have not come here at all.

I'll know who's signed as I go home; yes, pick out every
    one;
I'll watch you pass the drink saloon, and notice if you run.

So let me ask you once again to sign before you go;
For if you don't—mind, if you don't—Minnie is sure to
    know.

Appetite.

THOS. R. THOMPSON.

[For either a Girl or Boy.]

I'M going to speak of appetite; and when my piece is
    ended,
I hope you will not find a place you think could have been
    mended.

Now, appetite is right enough—a great and grand pos-
    session;
If natural entirely safe, and will the bones put flesh on.

If artificial, then look out! and pamper not the stranger;
An artificial appetite, if gratified, means danger.

An appetite for ale and beer, for brandy, gin, and whiskey,
Is artificial, and we know of all things, very risky.

All dangerous things we ought to shun, and so we think
    it better
To practice total abstinence unto the very letter.